FARMER LLAMA'S FARM MACHINES

PLANTERS

BY KIRSTY HOLMES

BEARPORT
PUBLISHING

Minneapolis, Minnesota

Fitchburg Public Library
5530 Lacy Road
Fitchburg, WI 53711

Library of Congress Cataloging-in-Publication Data
is available at www.loc.gov or upon request from
the publisher.

ISBN: 978-1-64747-545-1 (hardcover)
ISBN: 978-1-64747-552-9 (paperback)
ISBN: 978-1-64747-559-8 (ebook)

© 2021 Booklife Publishing
This edition is published by arrangement with
Booklife Publishing.

For more information, write to Bearport Publishing,
5357 Penn Avenue South, Minneapolis, MN 55419.
Printed in the United States of America.

IMAGE CREDITS

All images are courtesy of Shutterstock.com, unless otherwise specified. With thanks to Getty Images, Thinkstock Photo, and iStockphoto.
Cover - NotionPic, Tartila, A-R-T, logika600, BiterBig, studioworkstock. Aggie - NotionPic, Tartila. Grid - BiterBig. Farm - Faber14. 2 - movinglines.studio.
5 - Mascha Tace. 6 - Nucleartist. 7 - studioworkstock, David A Litman. 8 - movinglines.studio. 9 - Nucleartist, Ekaterina_Mikhaylova. 10 - Rimma Rii,
photowind. 11 - studioworkstock. 12 - NotionPic. 13 - Glukhova, Vasilyeva Larisa. 14&15 - Maike Hildebrandt. 14 - Niraelanor. 15 - K-Nick, StockSmartStart.
16 - studioworkstock. 17 - Kunturtle, ProStockStudio, JamesChen. 18&19 - Paul Kovaloff. 20 - Mascha Tace. 21 - DRogatnev. 22 - Dshnrgc, KASUE,
KenshiDesign, Piyawat Nandeenopparit. 23 - stockakia.

CONTENTS

DOWN ON THE FARM!

Welcome to Happy Valley Farm.
My name is Aggie. I'm a farmer llama.

Are you ready to lend a hoof?

If you've ever wondered how farmers grow so much food, you've come to the right place. I will teach you how!

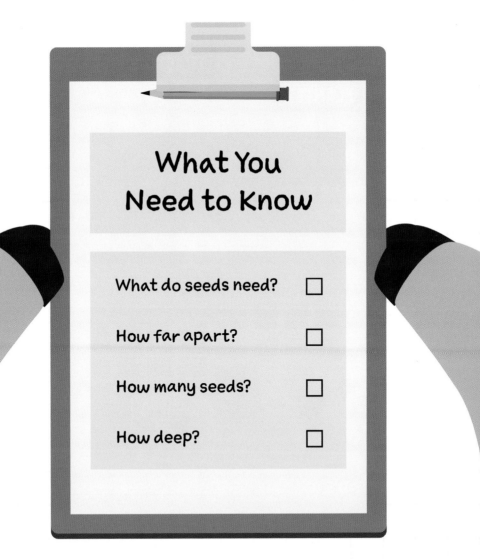

What You Need to Know

What do seeds need? ☐

How far apart? ☐

How many seeds? ☐

How deep? ☐

WHAT IS A PLANTER?

PLANTER

TRACTOR

A planter is a machine that plants **seeds** in the ground so they can grow. Planters are attached to tractors, which pull them along.

6

Crops are plants that are grown to eat. Most crops grow from seeds. Seeds planted in the ground grow into food.

BEFORE PLANTERS

Before machines, seeds had to be **sown** by hand. Farmers scattered seeds into the **furrows** they had made by plowing the field.

PLOWING TURNS THE SOIL AND LEAVES NEAT GROOVES IN IT.

This is called broadcast planting, because the farmer throws the seeds over a **broad** area.

Planting machines may have been used in ancient **Mesopotamia**. They were used in ancient China, too. The planter has been an important machine in human history.

Turn to page 12 to learn about the hopper!

HUMAN AND MACHINE

These are great for small fields or places a big tractor can't reach.

Some simple planters can be pushed along by hand. The hopper holding the seeds has little holes in it. As it is pushed along the ground, the seeds fall out.

Tractors can be used to pull big planters along. These machines can sow a lot of rows of seeds at once, cover them, and add **fertilizer** at the same time!

This saves farmers a lot of time and work.

PARTS OF A PLANTER

Let's look at the parts of a planter.

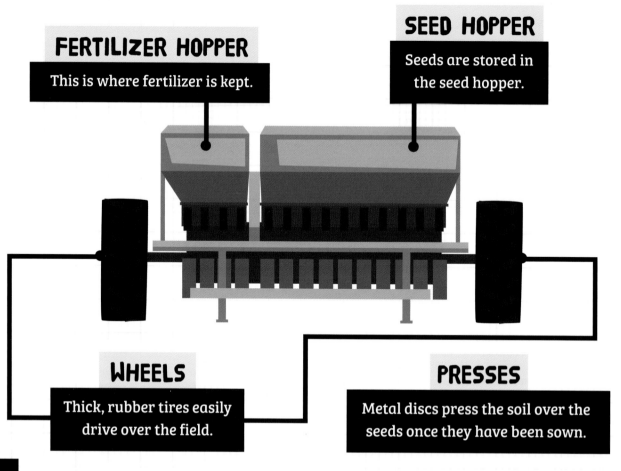

FERTILIZER HOPPER

This is where fertilizer is kept.

SEED HOPPER

Seeds are stored in the seed hopper.

WHEELS

Thick, rubber tires easily drive over the field.

PRESSES

Metal discs press the soil over the seeds once they have been sown.

Large, modern planters can have lots of parts. This one can turn the earth, plant the seed, add fertilizer, cover the seed, and pat it down, all in one pass!

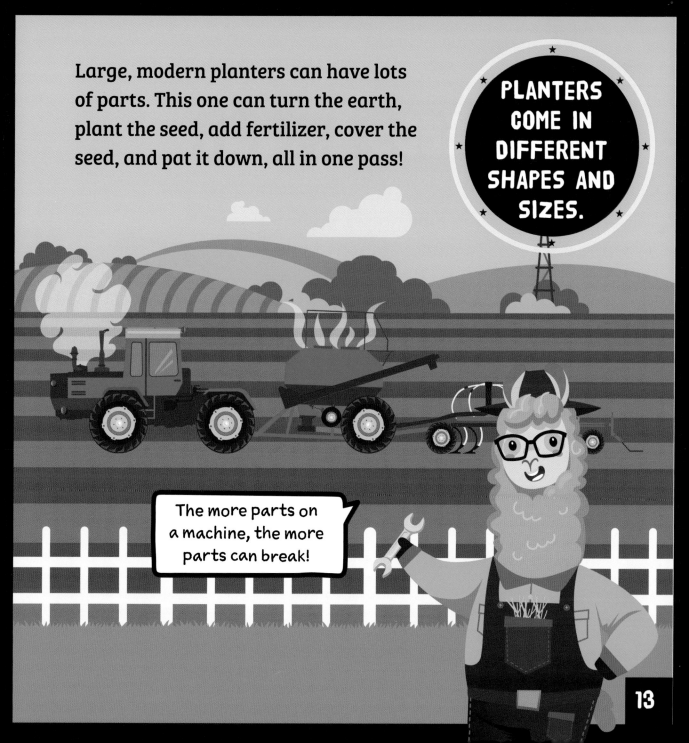

PLANTERS COME IN DIFFERENT SHAPES AND SIZES.

The more parts on a machine, the more parts can break!

WHAT SEEDS NEED

Seeds need sunlight, food, water, and warmth. Seeds are planted in soil at just the right **depth** so that these things can reach them.

TOO SHALLOW, AND BIRDS CAN EAT THE SEEDS. TOO DEEP, AND SUNLIGHT AND WATER WON'T REACH THEM.

Once the seed is planted at the right depth and has everything it needs, it will grow into a plant.

SEED DRILLS

I can tell the machine exactly how deep to plant these seeds.

Some planters are called seed drills. Seed drills push the seeds into the soil through small pipes to just the right depth.

Unlike a broadcast planter, this type of planter places seeds in neat rows. It can also change how far apart the seeds are.

SMALL LETTUCE DOESN'T NEED AS MUCH SPACE AS . . .

GIANT PUMPKINS!

RECORD BREAKERS

One of the biggest seed drills ever made weighed 33,000 pounds (15,000 kg) and was over 49 feet (15 m) long!

There are more than seven billion people on the planet. With so many people to feed, farms are getting bigger and bigger.

And so are the machines!

GET YOUR LLAMA-DIPLOMA

Well done! You made it through the training. If you've been paying attention, this test should be no prob-llama . . .

Questions

1. What is it called when a farmer sows seeds by scattering them by hand?

2. Name the things seeds need in order to grow.

3. What else can some planters add as well as seeds?

4. What can happen if a seed is too shallow?

5. What type of planter places seeds in neat rows?

You made that look easy! Welcome to the Happy Valley Farm family!

CERTIFICATE
OF APPRECIATION

LLAMA-DIPLOMA

Farmer Llama

Download your llama-diploma!

1. Go to **www.factsurfer.com**

2. Enter **"Planters"** into the search box.

3. Click on the cover of this book to see the available download.

KNOW YOUR BEANS

Of course, it's very important to know exactly what you are planting. Here at Happy Valley Farm, we grow one very special crop . . .

STEP ONE
Collect magic beans

STEP TWO
Grab fertilizer

STEP THREE
Wait for moonlight

GLOSSARY

BROAD covering a large or wide area

DEPTH the distance from the top to the bottom of something

FERTILIZER something that is added to soil that helps something grow

FURROWS long, narrow grooves made in the ground to plant seeds

MESOPOTAMIA a historical place that was where Iraq is today

SEEDS the parts of a flowering plant that can grow into new plants

SHALLOW not very deep

SOWN to have planted

INDEX